Drone Pre-Flight Checklist

For any commercial drone operation, it's crucial that you have a checklist template to run before each flight to make sure everything is in working order.

Safe operation and proper maintenance of your drone systems are paramount, and in order for you to ensure you're not putting your license, equipment, or the whole operation at risk, you simply can't afford to avoid a structured process.

If you're not taking proper steps to eliminate risk and uncertainty, you'll be wasting time and money trying to figure out what exactly went wrong.

What's more, it's incredibly important that pilots are properly compliant with safety regulations and proper procedure for safe, smart drone operation.

This checklist will run you through a comprehensive list of common problems and preventative measures you should be taking to ensure you're doing everything you can to mitigate risk and streamline your flight process.

Ultimately, this process can mean the difference between a successful operation and a fatal mission failure.

Weather

Weather can prove to be unpredictable and highly problematic for drone operations, so it's important to incorporate adequate checks into your process.

Especially at altitudes, it is crucial to be aware of local wind speed and the impact it could have on your flight.

A relatively simple check, but important nonetheless: taking the time to **assess whether or not visibility is consistent** throughout the expected range of operation.

Clouds, fog, and other environmental factors can appear quickly and significantly disrupt operations if they are not anticipated.

Precipitation can wreak havoc on drone systems, especially those with unnoticed structural damage and exposed electronics.

Assessing precipitation levels enables you to make more informed predictions about the most likely conditions for that day, and ultimately be more prepared to mitigate fault and error in your operations.

Best practice checks for weather include **consulting the UAV Forecast app or a local weather service provider.**

If you are operating close to an airport, you can also **check the local METAR / TAF readings**.

Ensure all conditions below are met before proceeding to the next task.

Is wind speed within an acceptable range?

☐ Yes

☐ No

Is visibility sufficient for the area of flight?

☐ Yes

☐ No

Is precipitation within an acceptable range?

☐ Yes

☐ No

Flight details

Record basic details

To begin with, you'll need to **record all the relevant identification of the pilot operating the drone.**

Equally, you will need to **record specific information about the drone unit in operation**

Use the form fields below to input the information.

- ☐ Pilot first name:
- ☐ Pilot last name:
- ☐ FAA / EASA registration number:
- ☐ Date of flight:
- ☐ Site location:
- ☐ Drone model name:
- ☐ Drone model number:
- ☐ Drone ID number (serial):
- ☐ Drone weight (kg):
- ☐ Drone repair technician email address:

State purpose of flight

Clear records of flight objectives are important for compliance, and a process which factors in this information can lead to useful datasets further down the line.

Clearly articulate the full scope of flight operations using the form fields below.

Flight purpose

- ☐ Recreation
- ☐ Commercial
- ☐ Search and rescue
- ☐ Other (please describe)

Does flight require access to restricted airspace?

- ☐ Yes
- ☐ No

Does flight require waiver for operational permission?

- ☐ Yes
- ☐ No

Describe alternative flight purpose

If you're not operating within the defined scope of the previous task, please **describe your flight purpose in the field below.** What is the purpose of the flight?

Check commercial drone license is valid

To fly unmanned air systems commercially in Europe, you need to **make sure your commercial drone license — otherwise known as a Remote Pilot Certificate — is valid and in-date.**

If you are flying as a government employee (police, fire, etc.) then you do not require a commercial drone license.

Check waiver for operational permission is valid

For some operations, **special permission will need to be granted, and a waiver or document of authorization for a particular regulation must be obtained from the National Aviation Autorithy.**

These non-airspace waivers include conditions such as flying at night, flying directly over a person or crowds of people, flying multiple aircraft with only one pilot, and flying above 400 feet (no possible yet)

Check that all waivers are correct for the needs of the operation and are in-date.

Check airspace authorization is valid

Airspace authorizations allow you to fly in otherwise restricted areas, like those governed by Air Traffic Control (ATC) regulations or other operations.

What's important here is that **your authorization covers the full scope of your flight plan** and that **the dates specified in the document are all valid for the mission at hand.**

Once you've verified the airspace authorization against your flight plan and made sure it's correctly dated you're good to go.

Batteries

Charge drone batteries

As you prepare your drone system for flight, you will need to **ensure all batteries contain enough charge for a successful payload.**

Maintaining a minimum of 75% charge per battery should help to reduce power-critical mission failure, and spares should be on hand for each unit to reduce time lost to charging.

Be sure to ensure all battery packs are securely fixed into position before launch.

- ☐ Charge drone battery to 75%

- ☐ Charge spares to 75%

- ☐ Securely fix drone battery into position

Charge controller batteries

Whether it's a portable ground control station or a traditional 'twin-stick' style handheld transmitter, **making sure all battery powered controller components are sufficiently charged** is paramount.

If your controller loses power, then you lose drone control.
- ☐ Charge controller battery to 75%
- ☐ Charge spares to 75%
- ☐ Secure controller batteries into position

Structural check

Inspect all components for visible damage

For any close inspection of your drone unit, especially involving propellers, you should ensure that the system is powered down completely before you begin.

Structural damage can appear anywhere on the drone unit in the form of small cracks or dings, and can go unnoticed without regular inspection. Over time and when compounded, these can seriously compromise the safety and proper functionality of your machine.

Inspect every component carefully for visual damage and **check off each one as you go.** After the whole unit has been inspected, **make a note if anything needs to be replaced or repaired.**

- [ ] Chassis
- [ ] Propellers
- [ ] Motors
- [ ] Gimbal
- [ ] Indicator lights
- [ ] Screws
- [ ] GPS
- [ ] Landing gear
- [ ] Batteries
- [ ] Electronic Speed Controller
- [ ] Compass
- [ ] Wiring
- [ ] Camera

Do any components need to be replaced or repaired?

- [ ] Yes
- [ ] No

Declare which components are in need of repair

If your drone is in need of repair, **state exactly what requires servicing below.**

Once you've specified repair work to be done, you'll want to **export or print this checklist and send it over to your maintenance technician.**

Which components need repair?

- [ ] Chassis
- [ ] Propellers
- [ ] Motors
- [ ] Gimbal
- [ ] Indicator lights
- [ ] Screws
- [ ] GPS
- [ ] Landing gear
- [ ] Batteries
- [ ] Electronic Speed Controller
- [ ] Compass
- [ ] Wiring
- [ ] Camera

Clear obstructions

Dirt and debris can cause parts of your system to malfunction or behave unexpectedly.

Checking and careful cleaning of specific areas should be performed after structural damage inspections.

Propellers should be able to move freely, camera lenses and covers should be checked for scratches and foreign objects, and status LEDs should be clearly visible. It's also possible for FOD (foreign object debris) to become trapped in enclosed spaces like the motor chamber.

Complete the sub-checklist below, confirming each component is unobstructed and clear of debris.

As with the previous task, the system should remain powered down throughout this check.

- ☐ Propellers
- ☐ Camera
- ☐ Status LEDs
- ☐ Motor frame

Secure additional components

With components cleared of obstruction and checked for structural damage, then the next step is to **fasten each component securely to the unit and double check the storage card is inserted correctly.**

Using the sub-checklist below, **make sure each component is securely in place.**

- ☐ Camera
- ☐ Gimbal
- ☐ Propellers
- ☐ Storage card
- ☐ Payload (if applicable)

Calibration

Check and update firmware

Firmware update checks should be performed before calibration to reduce unexpected behavior that might result from an updated component.

Drone firmware updates can be performed in two different ways; either by **connecting the drone to the computer directly with a USB or equivalent cable**, or **wirelessly via an application on a mobile device.**

Once you've updated firmware for both drone and controller units, **check both subtasks off below.**

- ☐ Drone firmware is up to date
- ☐ Controller firmware is up to date

Position antennas for signal strength

It is worthwhile spending time before launching to **position the antennas for best possible signal reception.**

Signal strength is too important for mission success to ignore, and antennas can often be nudged or displaced in transit or during a previous operation or routine check.

Once you've **repositioned the antennae and tested signal strength**, you'll be ready to calibrate the instruments without risk of inconsistency.

- ☐ Antennas positioned correctly
- ☐ Signal strength tested

Check system instrument calibration

Instruments should not be re-calibrated unless specifically observed to be working incorrectly. This is to reduce human error and likelihood of an improper re-calibration.

The compass and GPS instruments are crucial parts of your drone unit's intelligent navigation system. Most navigational features of the drone will depend on these sensors to work correctly, so it's important that you're deploying best practice for instrument recalibration.

Test the readings of the following components and **determine whether or not they require recalibration.**

Does the compass need calibrating?

☐ Yes

☐ No

Does GPS need calibrating?

☐ Yes

☐ No

Does RTH need calibrating?

☐ Yes

☐ No

Does IMU need calibrating?

☐ Yes

☐ No

Calibrate compass

Especially when changing locations, interference can cause problems with your compass.

Calibrating your compass is often as easy as accessing an in-app command and waiting for the software to do its job. **Make sure you consult your model-specific manual for the exact process.**

There are certain things that you should do before calibrating, **regardless of the model.** The following steps will ensure you reduce local interference and increase your chances of calibrating correctly.

Check off each of the subtasks below as you go.

☐ Remove all metal wearables (rings, watches)
☐ Ensure batteries have at least 75% charge
☐ Set the drone on a stable, level surface
☐ Perform calibration as directed by the manual

Calibrate Global Positioning System (GPS)

If your Global Positioning System (GPS) coordinates are not showing correctly, you will need to reset the component. This step requires an active internet connection (wired or wireless) so **be sure that you are connected before you start.**

Since GPS relies on external satellite systems to relay information, you will have to establish a secure connection to the signal to reset your geo-location coordinates.

This is usually done inside of the monitoring software/application that your drone uses. As always, **read the manual** to be sure.

Once you've reset the coordinates, you should also **check that they are correct** by cross-referencing against other GPS devices or even Google Maps.

- ☐ Reset GPS coordinates
- ☐ Check coordinates are correct

Calibrate Inertial Measurement Unit (IMU)

Your drone's Inertial Measurement Unit (IMU) is a system which uses a combination of accelerometers and gyroscopes, sometimes also magnetometers, to calculate angular velocity and linear acceleration.

Calibrating your IMU is simple - despite the complexity of these instruments, calibration is all done by software running on your drone or control station modules.

Before calibration, **make sure you have sufficient battery charge** for the whole process. **Follow the subtasks below** for best practice when calibrating your IMU.

- ☐ Check drone battery has at least 75% charge
- ☐ Set the drone on a stable, level surface
- ☐ Perform calibration as directed by the manual

Calibrate RTH (Return to Home)

The **RTH failsafe should be properly configured and tested** before a full-scale operation.

Depending on your system, the method will be different, but there are certain considerations that you should always take into account, such as **making sure the home coordinates are set to be higher than the highest potential obstacle in the immediate area.**

Testing your RTH coordinates against more than one satellite GPS reference is also a good idea for the sake of reliability.

Confirm that the following is true:

- ☐ RTH altitude is set above the highest point in the local area
- ☐ RTH coordinates have been input correctly

Set maximum flight altitude

Beyond RTH configuration, it is important to **set the general maximum altitude** to ensure you do not breach compliance limits or stray outside of the signal range of your controller.

Ensure that both of the following statements have been confirmed:

- ☐ Maximum flight altitude is within range of ground controller
- ☐ Maximum flight altitude within legal limit

Pre-launch

Position drone for launch

Now that almost everything has been checked, it's important to **observe the launch area and locate the most suitable launch site.**

Ideally, the drone will want to be set on an even surface, with nothing obstructing the area immediately above or around it.

Complete the subtasks below to ensure the area is clear for takeoff.

- ☐ Drone is on stable, level ground
- ☐ Immediate area around the drone is clear of obstructions
- ☐ Overhead is clear of obstructions

Check aircraft status LEDs

Finally, **pay attention to the status LEDs for signs of system errors or warnings.**

The LEDs should be indicating that everything is functioning normally - watch out for emergency or problem signals.

Refer to your unit's model-specific documentation for information about LED status signs and what they mean.

Reschedule flight date

If the day's weather means it won't be safe to operate the drone, you'll need to **schedule a new date for the flight.**

Choose a date in the form field below. You'll want to **check the forecast in advance leading** up to the new date.

☐ New flight date:

```


```

Made in the USA
Las Vegas, NV
29 September 2024

95965797R00017